AF132208

Environmental Science 101
Laboratory Manual

Spokane Falls Community College
Pullman Branch

Spring 2019

Linda Cook
Instructor

Copyright © 2019 by Linda Cook

Env Sci 101 Name:________________________________

Lab 1: The Metric System

The metric system is a simple and logical measurement system used in most countries throughout the world, and is used by scientists everywhere. In our labs this quarter we will be using the metric system, so this lab is designed to introduce you to this system and give you some practice using it.

If you can multiply and divide by 10 you can master the metric system. Larger and smaller units are found by multiplying or dividing by units of 10. This is the same as just moving the decimal point to the right or left.

Table 1.1 Metric prefixes

kilo = 1,000	milli = 1/1000
centi = 1/100	micro = 1/1,000,000

Table 1.2 Metric units of measure and their equivalents

Quantity	**Metric unit**	**Symbol**	**Equivalent**
Mass	Gram	g	1 g = 0.0022 pound
Length	Meter	m	1 m = 3.280 ft.
Volume	Liter	L	1 L = 1.057 qt
Temperature*	Degree Celsius	°C	1 °C = 1.898°F

*Because in the Fahrenheit scale freezing is 32°, you must *add* 32 when converting a °C measurement to °F.

Temperature measurements are simplified in the metric system because the freezing point of water is 0 °C, while the boiling point of water is 100 °C.

Complete the following sentences:

One kilometer is ____________________ meters, or ________________________ centimeters.

One ________________________ is 1,000 grams.

One liter is ______________________________ milliliters.

To convert from a larger unit to a smaller unit of 10, multiply by the correct unit or move the decimal the appropriate number of places to the right.

Example:
Convert 50 m to cm.

$$50 \text{ m} \cdot \frac{100 \text{ cm}}{\text{m}} = 5{,}000 \text{ cm}$$

50 x 100 = 5,000 cm

OR

Move the decimal two places to the right: 50.0 → 5000

To convert from a smaller unit to a larger unit, divide by the correct unit of 10 or move the decimal the appropriate number of places to the left.

> Example:
> Convert 50 m to km. $50\text{ m} \cdot \dfrac{\text{km}}{1000\text{ m}} = 0.05\text{ km}$
>
> 50/1000 = 0.05 km
>
> OR
>
> Move the decimal three places to the left: 50.0 → 0.05 km

Table 1.3 Metric to English and English to Metric conversion Tables

METRIC/ENGLISH CONVERSION FACTORS

ENGLISH TO METRIC	METRIC TO ENGLISH
LENGTH (APPROXIMATE)	**LENGTH** (APPROXIMATE)
1 inch (in) = 2.5 centimeters (cm)	1 millimeter (mm) = 0.04 inch (in)
1 foot (ft) = 30 centimeters (cm)	1 centimeter (cm) = 0.4 inch (in)
1 yard (yd) = 0.9 meter (m)	1 meter (m) = 3.3 feet (ft)
1 mile (mi) = 1.6 kilometers (km)	1 meter (m) = 1.1 yards (yd)
	1 kilometer (km) = 0.6 mile (mi)
AREA (APPROXIMATE)	**AREA** (APPROXIMATE)
1 square inch (sq in, in^2) = 6.5 square centimeters (cm^2)	1 square centimeter (cm^2) = 0.16 square inch (sq in, in^2)
1 square foot (sq ft, ft^2) = 0.09 square meter (m^2)	1 square meter (m^2) = 1.2 square yards (sq yd, yd^2)
1 square yard (sq yd, yd^2) = 0.8 square meter (m^2)	1 square kilometer (km^2) = 0.4 square mile (sq mi, mi^2)
1 square mile (sq mi, mi^2) = 2.6 square kilometers (km^2)	10,000 square meters (m^2) = 1 hectare (ha) = 2.5 acres
1 acre = 0.4 hectare (he) = 4,000 square meters (m^2)	
MASS - WEIGHT (APPROXIMATE)	**MASS - WEIGHT** (APPROXIMATE)
1 ounce (oz) = 28 grams (gm)	1 gram (gm) = 0.036 ounce (oz)
1 pound (lb) = 0.45 kilogram (kg)	1 kilogram (kg) = 2.2 pounds (lb)
1 short ton = 2,000 pounds (lb) = 0.9 tonne (t)	1 tonne (t) = 1,000 kilograms (kg)
	= 1.1 short tons
VOLUME (APPROXIMATE)	**VOLUME** (APPROXIMATE)
1 teaspoon (tsp) = 5 milliliters (ml)	1 milliliter (ml) = 0.03 fluid ounce (fl oz)
1 tablespoon (tbsp) = 15 milliliters (ml)	1 liter (l) = 2.1 pints (pt)
1 fluid ounce (fl oz) = 30 milliliters (ml)	1 liter (l) = 1.06 quarts (qt)
1 cup (c) = 0.24 liter (l)	1 liter (l) = 0.26 gallon (gal)
1 pint (pt) = 0.47 liter (l)	
1 quart (qt) = 0.96 liter (l)	
1 gallon (gal) = 3.8 liters (l)	
1 cubic foot (cu ft, ft^3) = 0.03 cubic meter (m^3)	1 cubic meter (m^3) = 36 cubic feet (cu ft, ft^3)
1 cubic yard (cu yd, yd^3) = 0.76 cubic meter (m^3)	1 cubic meter (m^3) = 1.3 cubic yards (cu yd, yd^3)
TEMPERATURE (EXACT)	**TEMPERATURE** (EXACT)
[(x-32)(5/9)] F = y C	[(9/5) y + 32] C = x F

QUICK INCH - CENTIMETER LENGTH CONVERSION

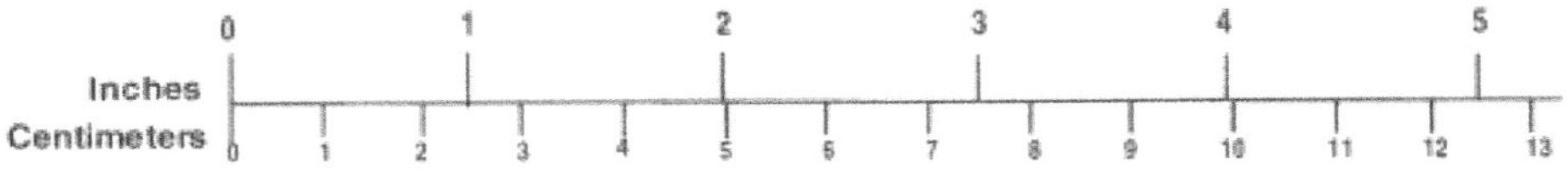

QUICK FAHRENHEIT - CELSIUS TEMPERATURE CONVERSION

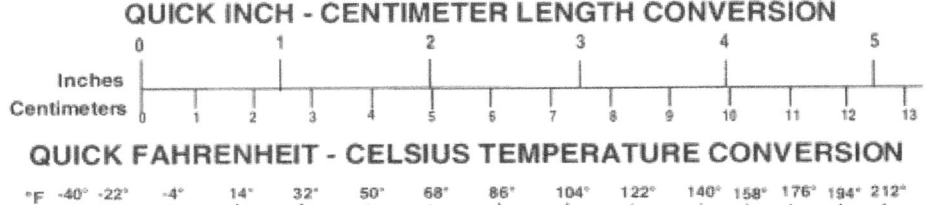

°F	-40°	-22°	-4°	14°	32°	50°	68°	86°	104°	122°	140°	158°	176°	194°	212°
°C	-40°	-30°	-20°	-10°	0°	10°	20°	30°	40°	50°	60°	70°	80°	90°	100°

For more exact and or other conversion factors, see NIST Miscellaneous Publication 286, Units of Weights and Measures. Price $2.50 SD Catalog No. C13 10286

Updated [illegible]

Using the conversion tables above, answer the following questions.

1. How many centimeters long is a 3.5 inch fish?

2. How many miles per hour can you drive if the speed limit is 105 km/h?

3. There are two gas stations on either side of the Canadian border. Which country has cheaper gasoline if Canada's price is $0.38 a liter and the U.S. price is $1.30 a gallon (prices are in USD)?

4. The recipe for Swiss chocolate mousse calls for 12.5 ml of salt. How many tsp. of salt will you put in?

5. You know that trout prefer water temperatures between 50° and 60°F. You measure the water temperature of a stream and find that it is 13°C. Would this be a good place to sample trout populations?

6. Could you carry a 70 kg specimen back to the lab?

REVIEW: Table 1.4 shows the most commonly used metric measurements. You will use many of these units of measure during lab activities. Apply the concepts you've learned so far in lab to complete the table. Try to do it without peeking back in the lab!

Table 1.4

Length:

Meter

1 centimeter = ___________ m

__________ mm = 0.001 m

1 km = _______________ m

Mass:

Gram

1 milligram = ______________ g

___________ kg = 1,000 g

Volume:

Liter

1 milliliter = _________ L

Temperature:

$^{\circ}$ Celsius

Water freezes at ______ $^{\circ}$C and boils at ______ $^{\circ}$C.

EXERCISE:

Measure the following and record your measurements.

I. Length station

a. Measure your lab partners' height using a meter stick or metric tape measure. Record his/her height in cm and in m.

Lab partner's height: ______________ cm ___________________ m

Lab partner's height: ______________ cm ___________________ m

Lab partner's height: ______________ cm ___________________ m

Your height: ______________ cm ___________________ m

Answers to calculation on previous page

A1.	3.5 in. x 2.54 cm/in = 8.89 cm
A2.	105 km/h x .62 mi/km = 65mi/h
A3.	U.S. (1 gal = 3.79 L so \$0.38 x 3.79 = \$1.44/gallon)
A4.	12.5 x 0.2 = 2.5 tsp.
A5.	(13 x 1.8) + 32 = 55°F This is optimal temperature for trout.
A6.	70 kg x 2.2 pounds/kg = 154 pounds. How far is it to the lab?

b. Measure and record the following in cm:
   i. Length of your textbook: ________________cm

   ii. Length, height, and width of a shoebox: ______cm, _______ cm, _______cm

   iii. Now calculate the volume of the shoebox: _____________cm³

II. Volume station

a. In the metric system, a ml = 1 cc (cubic centimeter, or cm³). What is the volume in ml of one white cube? ___________ml

b. What is the volume in ml of one blue rod? _________ ml

c. An irregular object's volume can be determined by the volume of water it displaces. Record the volume of water in the graduated cylinder (read the volume at the **bottom** of the meniscus); then drop the rock labeled #1 into the water (carefully!). Record the new volume.

Volume of water + rock in ml ___________in cc__________

Volume of water originally in ml _________ in cc_________

Difference in ml ___________ in cc__________

The difference is the volume of the rock.

d. Now predict (or estimate) what you guess the approximate volumes of the other three rocks to be:

Rock #2: _____________

Rock #3: _____________

Rock #4: _____________

e. Complete the table below:

Table 1.5

	Volume rock + water (ml)	Original volume water (ml)	Volume of rock (ml)
Rock #2			
Rock #3			
Rock #4			

f. You notice there are two sizes of graduated cylinders. Not that they are marked in different increments. How did you decide which cylinder to use for which rock?

What advantage is there to using the smaller cylinder for measuring the smaller rocks and the larger cylinder for the larger rocks? What disadvantage might there be to this method?

Finally, what advantage might there be to using the larger cylinder for measuring all the rocks? What disadvantage?

III. Mass station

a. Step on the scale and record your mass in kg: ____________________

b. Convert to pounds: ______________________________

IV. Temperature station

a. Normal body temperature in degrees F is 98.6. Convert this to normal body temperature in degrees Celsius: ____________

b. Close your hand around the thermometer and record the temperature of your hand: ________________

Enter all the data from the following exercises in the class-wide tables provided on the following page:

Ia, IIIa, and IVb.

Student	Height (cm)
1	
2	
3	
4	
5	
6	
7	
8	
9	
10	
11	
12	
13	
14	
15	
16	
17	
18	
19	
20	
21	
22	
23	
24	
25	

Student	Mass (kg)	Palm temp (°C)
1		
2		
3		
4		
5		
6		
7		
8		
9		
10		
11		
12		
13		
14		
15		
16		
17		
18		
19		
20		
21		
22		
23		
24		
25		

Env Sci 101 Name:________________________________

Lab 2: Graphing

Experimental data are often shown in the form of a graph, a diagram showing the relationship between the **dependent** and **independent** variables. The dependent variable is usually shown on the y-axis (vertical axis) and the independent variable is shown on the x-axis (the horizontal axis). A graph can help to visualize and interpret data more clearly and easily than a table.

When you are creating a graph, keep in mind that your goal is to show data in the clearest, most readable form possible. Always use graph paper for graphs in this class (or you may use a computer program such as Excel if you know how). The intervals on each axis should be appropriate for the range of data such that most of the graph is used. The intervals on the axes must be equal. Label each axis with the name of the variable and the units in which it is measured. Your graphs should be titled and may also include an explanatory caption if necessary.

The following table tracks the growth of a population of bacteria in a lab. Under some conditions bacteria are capable of dividing to produce two new cells from each parent cell every 20 minutes. Assume that 3 bacteria are placed in a flask that is kept at ideal conditions of temperature and light and contains adequate nutrients and oxygen to sustain bacterial growth for two days.

Table 2-1 Population Growth in Bacteria

Time (minutes)	**Population size**
20	3
40	6
60	
80	
100	
120	
140	
160	
180	
200	
220	
240	

1. Fill in Table 2-1 above.

2. Graph population size (y-axis) against time (x-axis) below. Remember to label appropriately!

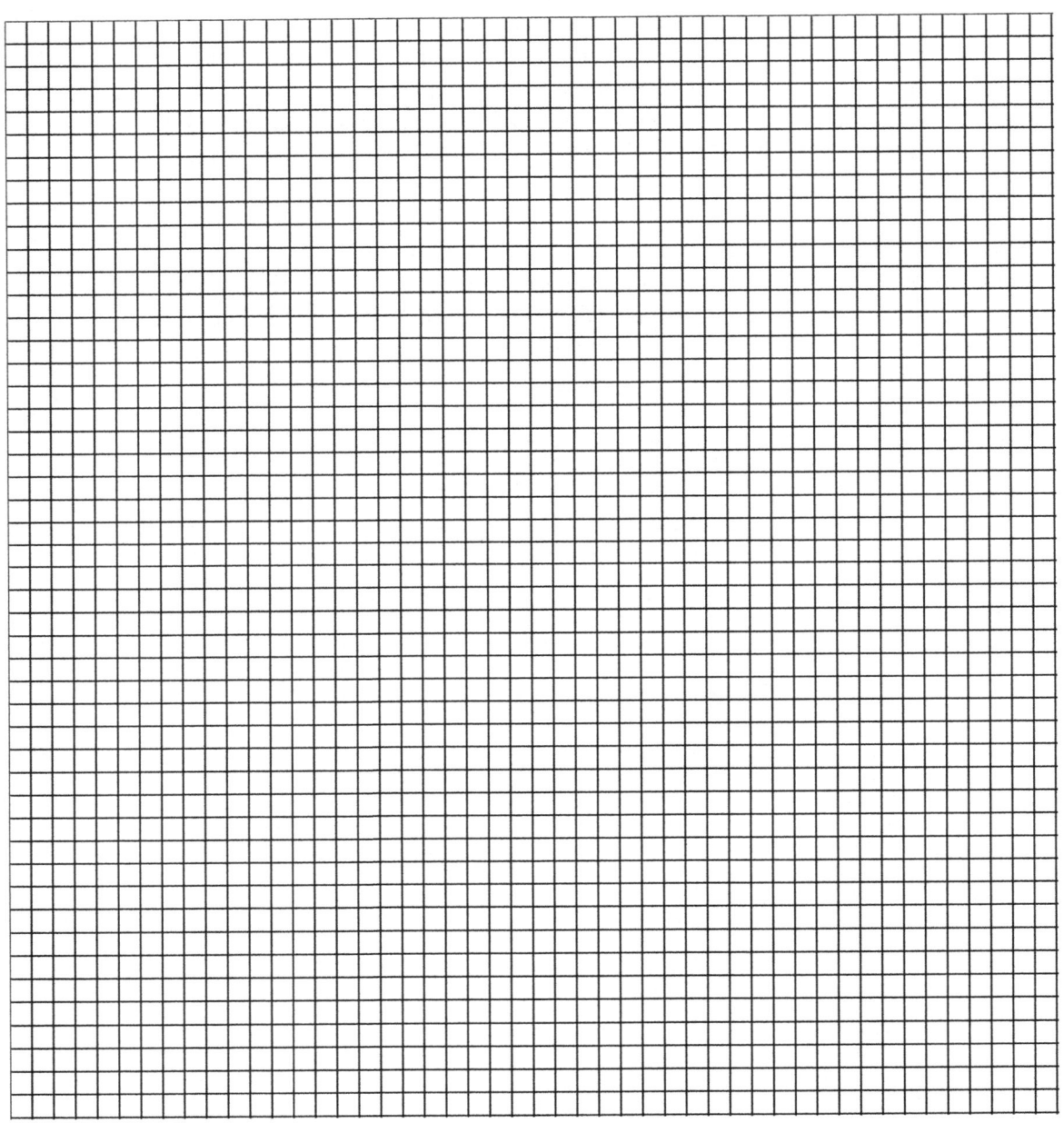

3. Use the tables of data provided for the first laboratory. Plot the height data for the class as a simple distribution (round your measurements to the nearest 5 cm). Place x's at the appropriate points on the number line to create a distribution curve for these data.

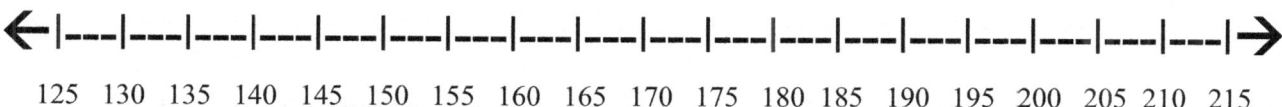

←|---|---|---|---|---|---|---|---|---|---|---|---|---|---|---|---|---|---|→

125 130 135 140 145 150 155 160 165 170 175 180 185 190 195 200 205 210 215

Height in cm

Turn the page; you are not done yet!

4. You have data on both body mass and the temperature of the hands for the individuals in the class.
   a. Do you think there might be any relationship between a person's body mass and the temperature of their extremities?

   b. Write a hypothesis about the relationship (if any) between these variables.

   c. Plot the body mass data from the class against the temperature data from our hands below. Label appropriately.

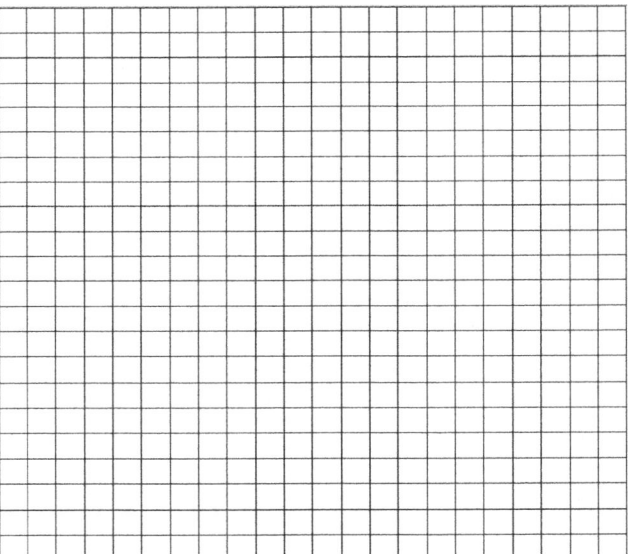

   d. Does there appear to be any relationship between these two variables? If so, what? Do the data appear to support your hypothesis or not?

5. The amount of oxygen dissolved in water is important to the organisms that live in a river. The amount of dissolved oxygen varies with changes in both physical factors and biological processes. The temperature of the water is one physical factor affecting dissolved oxygen levels as shown in the data table below. The amount of dissolved oxygen is expressed in parts per million (ppm).

Dissolved Oxygen Levels at Various Water Temperatures

Water Temperature (°C)	**Level Dissolved Oxygen (ppm)**
1	14
10	11
15	10
20	9
25	8
30	7

Using the information in the table, construct a line graph of these data.

a. Mark an appropriate scale on each axis.
b. Label the axes and title your graph.
c. Plot the data for dissolved oxygen on the grid. Each piece of data is a point. Connect the points with lines.

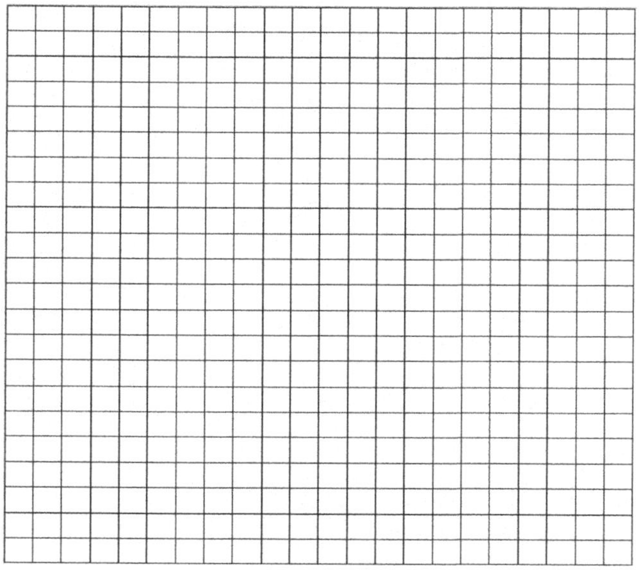

d. If the trend continues as shown in the data, what do you predict will be the dissolved oxygen level at a water temperature of 35°C?

e. What is the relationship between dissolved oxygen levels and water temperature?

6. A number of sunflower seeds planted at the same time produced plants that were later divided into two groups, A and B. Each plant in Group A was treated with gibberellic acid (a plant hormone) while the plants in Group B received no treatment. All other growth conditions were kept constant. The height of each plant was measured on five consecutive days, and the average height for each group recorded in the table below.

Average Plant Height (cm)

	Day 1	Day 2	Day 3	Day 4	Day 5
Group A	4.5	6	9	12	14
Group B	4.5	5.5	6	6.5	7

Using these data, construct a graph showing the growth curves for both sets of plants on the grid below. Remember to label everything (some of the labels are there but remember you need to indicate the scale on the y-axis). Use the key indicated to differentiate between the two groups of plants.

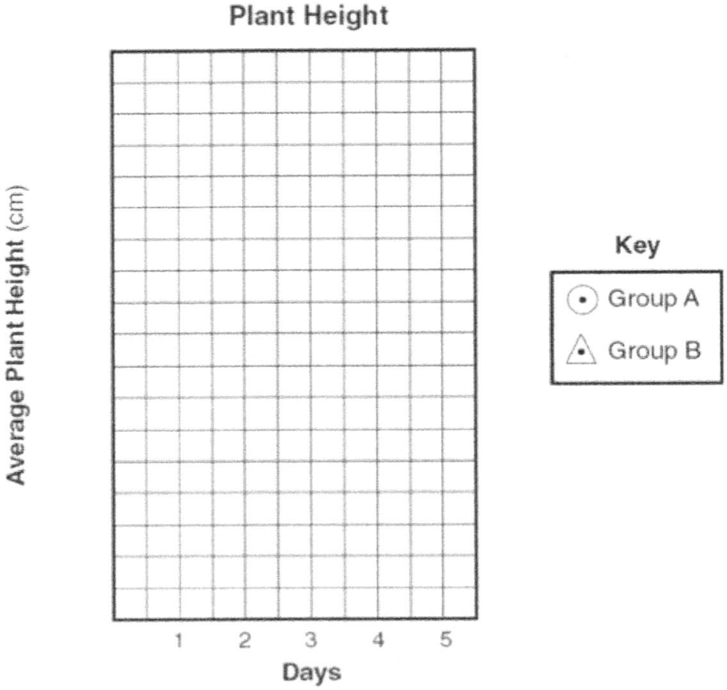

a. The dependent variable in this investigation was
    a. Days
    b. Plant height
    c. Gibberellic acid
    d. Group B

b. State a conclusion that can be drawn from these data about the effect of gibberellic acid on plant growth.

Turn the page; you are not done yet!

7. Each year, a New York State power agency provides its customers with information about some of the fuel sources used in generating its electricity. The table below shows this information from 2002-2003.

Fuel Sources Used

Fuel Source	**Percentage of Electricity Generated**
Hydro (water)	86
Coal	5
Nuclear	4
Oil	1
Solar	0

Mark an appropriate scale on the axis labeled "Percentage of Electricity Generated." Construct vertical bars to represent the data. Shade in each bar.

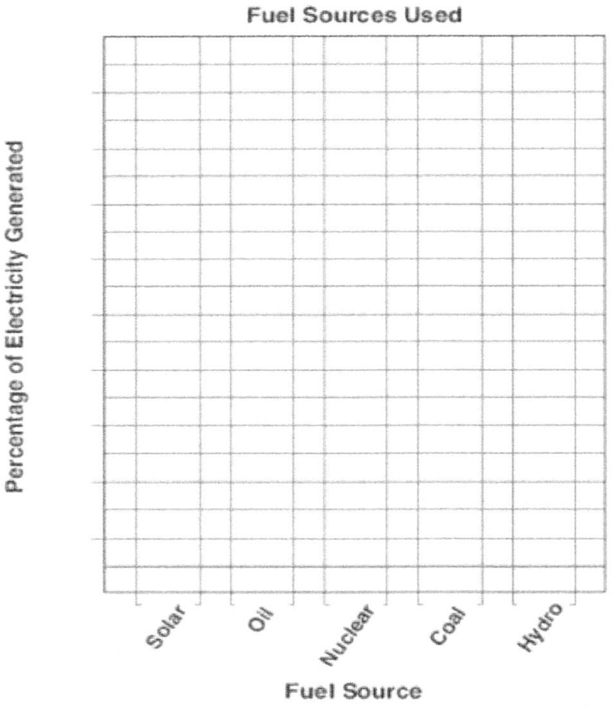

a. Which of the fuel sources used are considered fossil fuels?

b. Identify the sources in the table that are classified as renewable.

8. In the mid-1990s, for the first time in 70 years, the howl of wolves was heard in Yellowstone National Park. *Canis lupus*, the gray wolf, one of the largest and most complex canines and an apex predator, was successfully reintroduced to its historic territory in Yellowstone. In mid-January 1995, 14 wolves from many separate packs were captured in Canada and transported into Yellowstone. After being penned in three one-acre pens to allow them to develop a pack structure, they were released into the park. For the next four years the number of pups in the pack was censused. These data are shown below.

Number of Wolf Pups Observed

Year	Number of Pups
1996	11
1997	64
1998	42
1999	61

On the grid below, graph these data however you like (bar graph or line graph). Do not forget a title and all appropriate labels!

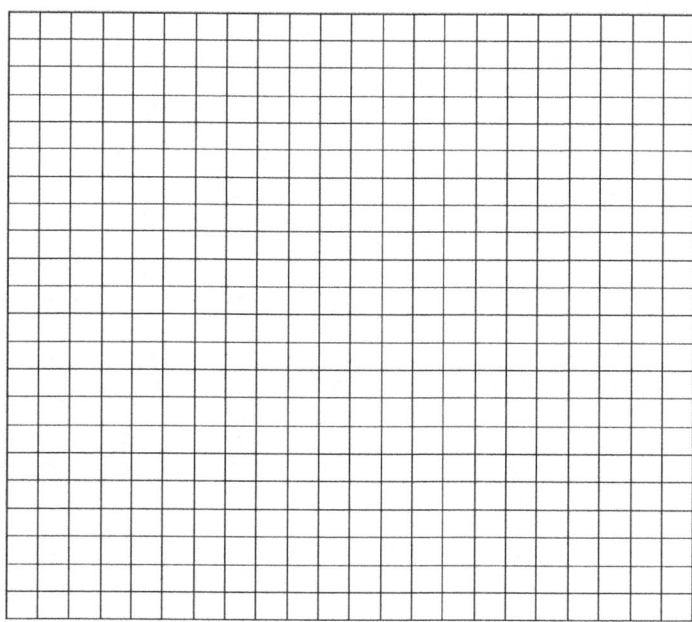

a. What is a possible reason for the decline in the number of pups in the third year of the study?

Env Sci 101 Name:________________________________

Lab 3: Plagiarism and Citations

1. In what two situations do you NOT need to provide a source for information in your paper?

2. When you have two or more consecutive sentences from the same source, in which sentence(s) does the citation belong? And what else needs to be done to indicate the source for the entire passage?

3. True or False: Direct quotes must be used sparingly, only when it is absolutely necessary to retain the exact wording of the original source.

4. Explain what an in-text citation is and how it should be formatted for APA-style.

5. Format the following information so it conforms to APA style for a reference.

The American Journal of Medicine
2013, volume 126, number 7, pages 583-589
Article: The Impact of Marijuana Use on Glucose, Insulin, and Insulin Resistance Among U.S. Adults.
Authors: Elizabeth A. Penner, M.D., Hannah Buettner, B.A., Murray A. Mittleman, M.D.

Please read the passage from the original source below. Then read Statements 6-10, which are passages in fictitious student papers. For each passage (Statements 5-9), indicate whether or not plagiarism has occurred, and explain why or why not.

The Original Source (Edward O. Wilson, The Diversity of Life, published 1992, pg. 164):
The most potent keystone species known in the world may be the sea otter. This wonderful animal, large and supple in body, cousin to the weasels, whiskered like a cat, staring with a languorously deadpan expression, once thrived among the kelp beds close to shore from Alaska to southern California.

Statement 6: The most potent keystone species known in the world may be the sea otter, which once thrived among the kelp beds close to shore from Alaska to southern California.

Statement 7: The most potent keystone species known in the world may be the sea otter (Wilson, 1992). Wilson explains that this wonderful animal, large and supple in body, cousin to the weasels, whiskered like a cat, staring with a languorously deadpan expression, once thrived among the kelp beds close to shore from Alaska to southern California.

Statement 8: In his book on biological diversity, Wilson (1992) described the sea otter in part as being "whiskered like a cat, staring with a languorously deadpan expression" (pg. 164).

Statement 9: Because of the unique relationship between the sea otter and the kelp beds and other organisms living along the North American Pacific Coast, it is considered by some ecologists to be a very important keystone species (Wilson, 1992).

Statement 10: According to Wilson (1992), the most important keystone species in the world might possibly be the sea otter. This magnificent animal, a close relative to the weasels, once lived among the kelp beds near the shore from Alaska to southern California.

Plant Evolution Timeline

START

4.6 billion years ago

- 3.6 billion years ago: 1st simple cells
- 3.4 billion years ago: 1st photosynthetic bacteria
- ~2 billion years ago: 1st eukaryotic cells
- ~1.2 billion years ago: 1st multicecllular organisms

BRYOPHYTES

- ~400 mya: 1st *land* plants
- Reproduce via spores (no seeds)
- Mosses, liverworts, hornworts
- No vascular tissue, so no true roots or leaves

FERNS & FRIENDS

- ~360 mya: 1st *vascular* plants:
- horsetails, club mosses, ferns,
  spike mosses
- phloem and xylem in true roots and leaves
- Reproduce via spores
  (no seeds)
- Spores borne on sporophylls

1ST GYMNOSPERMS (SEED PLANTS)

- ~300 mya
- Cycads, conifers, gnetophytes, & *Ginkgo*
- Cycads
  - Dioecious plants
  - Seeds produced in megastrobili (female cones)
- Conifers
  - Monoecious plants
  - Pines, spruce, firs, sequoia, larches, yews
  - Leaves needlelike or scalelike
  - Almost all are evergreen

FLOWERING PLANTS (ANGIOSPERMS)

- ~130 mya
- **Flowers**: modified clusters of leaves (strobili)
- Ova and pollen grains borne in modified sporophylls
- Fruits are mature ovaries of flower

Env Sci 101 Name:________________________________

Lab 4: Plant Anatomy

Welcome to WSU's School of Biological Sciences teaching greenhouse!

Today you are going to apply some of the plant anatomy that you've been learning in the classroom, and learn a little more. You will have a practical quiz covering this material next week so take your time and ask questions if you are not sure about something.

1.) Find the plant tagged with the letter **A**.
This plant is native to hot, dry areas with little rainfall and no trees. What adaptations do you note on this plant that suit it for that habitat?

Describe this plant's leaves.

2.) Find the plant tagged **B**. Look at the soil this plant is growing in (the big length of clay pipe contains the plant's long taproot). What can you hypothesize about this plant's native habitat?

3.) Find the plant tagged **C**. Sketch a *single leaf* from this plant in the box to the right. From the list below, circle as many terms as you can that describe this leaf.

Simple leaves	
Pinnately compound leaves	Once-compound
Palmately compound leaves	Twice-compound
Pinnate leaflets	Thrice-compound
Palmate leaflets	Petiolate
Sheathing leaf bases	

Plant C

4.) Find the plant with the **D** tag. Is this plant a monocot or a dicot? How do you know?

5.) Find the plant with the **E** tag. Identify the curly appendages at the tips of its branches. What are these called (botanically) and what is their purpose? Also, what part of the plant are they (root, stem, or leaf)?

6.) Find the plant with the **F** tag.

a. To which of the major plant groups we discussed does this plant belong?

____

b. Does this plant produce seeds? YES/NO

c. Does this plant produce flowers? YES/NO

d. Is this plant monoecious or dioecious?______________________________

7.) Find the plant with the **G** tag. **Please do not touch this plant. ☺

a. **Sketch** a *single leaf* from this plant in the box to the right.

b. From the list below, **circle** the terms that describe this plant.

Pinnately compound leaves	Once-compound
Palmately compound leaves	Twice-compound
Pinnate leaflets	Thrice-compound
Palmate leaflets	

Plant G

8.) Find the plant with the **H** tag.
From the list below, **circle** the terms that describe this plant.

Alternate leaf arrangement	
Opposite leaf arrangement	
Whorled leaf arrangement	Sheathing leaf bases
Simple leaves	Petiolate leaves
Compound leaves	

9.) Find the plant with the **I** tag.

a. Sketch a single leaf from this plant in the box below.

b. On your sketch, circle and label a single leaflet.

c. Turn over one leaf of this plant. What are those little circular bumps on its underside? What are they for?

Plant I

10.) Find the plant with the **J** tag. Examine this plant's flowers (be careful not to knock them off! But if some have fallen onto the bench you may pick them up to examine them).

a. Are they (the flowers) monoecious or dioecious?

b. *If* they are dioecious, can you tell if they are male or female? How do you know?

c. Is the *plant* monoecious or dioecious?

11.) Find the plant with the **K** tag.

a. Is this plant monoecious or dioecious? How do you know?

b. What evolutionary innovation is represented by the yellow-and-white showy structures on this plant?

12.) Find the plant with the **L** tag.

a. Describe the leaf bases you see on this plant.
b. Describe the venation of its leaves.

13.) Find the plant with the M tag.
From the list below, **circle** the terms that describe this plant.

Alternate leaf arrangement
Opposite leaf arrangement
Whorled leaf arrangement
Simple leaves
Compound leaves

14.) Find the plant with the **N** tag.
From the list below, **circle** the terms that describe this plant.

Alternate leaf arrangement	Simple leaves
Opposite leaf arrangement	Compound leaves
Whorled leaf arrangement	Sheathing leaf bases
	Petiolate leaves

15.) Find the plant with the **O** tag.

a. Describe the leaves of this plant (e.g., arrangement on stem, compound vs. simple, etc.).

b. These leaves are covered with specialized fuzzy hairs called trichomes. What do you hypothesize is their purpose?

16.) Find the plant with the **P** tag. Note that this plant is climbing up and around the trunk of another plant. Identify the blackish projections arising from the nodes of the stem of the climbing plant; what do you think these projections might be? What is their purpose?

Env Sci 101 Name:______________________________

Lab 5: Evolution and Adaptations

Adapted from Washington State University's Biol 102 Laboratory Manual, 2015 edition

There are about 1.5 million species known in the world, but scientists estimate that there may be as many as 30 million species in the world. This means we have identified only a tiny part of the diversity that exists on Earth. For every ecosystem on the planet there are any number of ecological niches, and organisms that are specifically adapted to live in these niches. Body shape, coloration, fur or feathers or scales, jaw or beak shape—these and many more are characteristics that may be adapted to help an organism survive.

Types of adaptations:

A. **Coloration**
   a. **Cryptic coloration**:
      i. Camouflage = matching the background; blending in with surroundings
      ii. Countershading = darker coloration on top and lighter coloration underneath can make an animal's shadow less apparent, and make them harder for a predator to see
      iii. Disruptive coloration = bright stripes and contrasting colors may break up an animal's outline, making it harder for predators to target individuals
   b. **Warning coloration**: animals that are poisonous or taste bad are often brightly colored as a signal to predators. Animals that are not poisonous but have warning coloration are called mimics; they have evolved to imitate the toxic prey animals as a protection.
   c. **Epigamic coloration**: coloration used to attract a mate. Usually such coloration is bright and flashy. This helps animals identify their own species, as well as advertise their strength and health to potential mates.

B. **Specializations for feeding**: animals exhibit different morphologies of head, teeth, and jaws depending on what and how they eat. Carnivores, for example, have very strong jaws and often pointed or serrated teeth for ripping and chewing, while herbivores may have broad flat teeth for grinding, and their jaws may be modified to chew in a different motion. Bird beaks vary widely with their diets (see examples throughout room 126).

C. **Visual adaptations**: aside from the eye itself (which has been adapted and modified in thousands of ways), the placement of eyes on an animal's head reveals much about the animal's lifestyle. Using birds as an example:
   a. Owls have eyes both facing forward from a flattened face. This gives them excellent binocular vision, depth perception, and ability to gauge the speed at which prey is moving. To see to the sides or behind them they must turn their heads.
   b. Sparrows and other songbirds' eyes are set further back on the sides of their heads. They do have some binocular vision but must also have a wide peripheral field of vision to watch for predators.
   c. Shorebirds hunt mainly by feel, probing with their bills into water or sediment. So it is more advantageous to them to see what is to the sides and behind, instead of focus in front of them. Their eyes are set far back on the sides of their heads so

they do not have binocular vision in the front, but can see in a complete circle without turning their heads.

**Can you think of examples of similar adaptations for vision in mammals?

D. **Limb adaptations**: The wing of a bird is a modified forelimb (in humans it would be an arm; in dogs or horses a front leg, and in dolphins or manatees a front flipper). Look in the hallway between rooms 135 and 138 to see other limb adaptations.
E. **Adaptations for flight**: In addition to wings, birds have other special features for flight. Hollow, light bones allow them to leave the ground. Feathers provide lightweight insulation and act as airfoils, as well as smoothing the body to reduce air resistance.
F. **Body shape**: Animals that live or hunt in burrows are often long and narrow in body, with short legs and small ears. Fliers and swimmers often have streamlined body shapes to lower resistance as they move through air or water.
G. **Adaptations for maintaining body temperature**: Cold-climate animals often have heavy thick coats of fur or feathers, and perhaps thick layers of fat, to serve as insulation. Animals from cold climates have reduced surface area (small ears, compact bodies) to reduce heat loss, while animals from hot climates have often increased surface area (large ears, longer bodies and tails) through which heat can leave their bodies.
H. **Dimorphism**: means "two forms." Species may exhibit sexual dimorphism where the males and females vary in color, size, etc. There is frequently age dimorphism where young appear different from adult forms. And there can be seasonal dimorphism, where animals change between summer and winter (coat color, presence or absence of antlers…).

All these are types of adaptations you will be able to observe in the Connor Museum of Zoology for this lab, as you complete the following exercises.

As you observe the animals in the museum, think about how each is adapted to its particular environment and lifestyle. **Use the list of adaptation concepts you have been given** to describe these adaptations for each of the animals listed below. When you have filled out all the answers, hand in your work.

Example: Aardvark (Located room 126)
Habitat: South Africa; plains, grasslands; feeds on termites and ants
Special adaptations: long snout for reaching into anthills and termite mounds; long thick nails for digging; sticky tongue for catching food; light-colored fur reflects heat

1. **Chinook salmon** (room 126)
   Habitat: cold freshwater streams as juveniles (fry); open ocean as subadults and adults; predatory (feeds on smaller fish, insects)
   Special adaptations:

2. **Prosaurolophus** (room 126)
   Habitat: warm, wet marshes and forests
   Special adaptations:

3. **Kangaroo** (room 126)
   Habitat: Australian grasslands, often dry habitats
   Special adaptations:

4. **Coyote** (room 138)
   Habitat: prairies; forests; urban areas
   Special adaptations:

5. **Ptarmigan** (hall display)
Habitat: tundra or mountains; eat seeds, buds of trees
Special adaptations:

6. **Cougar** (room 138)
Habitat: mountains, woody areas; mostly active at night; apex predator
Special adaptations:

7. **Pileated woodpecker** (room 135)
Habitat: mature forests, feeds on insects in tree bark
Special adaptations:

8. **Rabbit** (room 138)
Habitat: many habitats (woods, plains, mountains…); herbivores feeding on grasses, vegetation
Special adaptations:

9. **Porcupine** (room 138)
Habitat: forested areas; eat leaves, fruit, bark, nuts, berries, flowers; preyed upon by large predators such as coyotes
Special adaptations:

10. **Townsend's mole** (room 138)
Habitat: western Washington meadows, yards, lawns and gardens
Special adaptations:

11. **Hairy armadillo** (room 138)
Habitat: Brazilian fields, grasslands; nocturnal; eats small invertebrates (insects); considered a pest in parts of the US because they dig burrows in yards and gardens to find grubs to eat
Special adaptations:

12. **Beaver** (room 138)
Habitat: streams and lakes in woods, mountains; herbivores, eat herbaceous plants and bark; live in burrows or underwater dens/lodges they construct from branches
Special adaptations:

13. Find the display on lead poisoning in waterfowl.
    a. How do waterfowl get lead poisoning?

    b. What impact does lead poisoning have on these birds?

    c. What solutions have been suggested for this problem? Can you think of additional possible solutions?

14. Find the case containing the accipiters (room 135).
    a. What adaptations do all these birds share? What lifestyle features do they all have in common?

15. Find the case with the owls (room 135).
    a. What adaptations do all these birds share?

    b. How do these birds differ from the accipiters?

16. Find the display on adaptations of bird hindlimbs (room 126). Give at least four (4) examples of specialized adaptations in bird hindlimbs that are presented in this display. Now find one more example from somewhere else in the museum and describe it.

17. Find the display on behavioral adaptations in mountain goats (room 135) and describe or summarize one way in which these animals exhibit adaptive behavior.

Env Sci 101 Name:_______________________

Lab 6: Population Sampling

Adapted from Carolina EcoKits Simulating Methods to Estimate Population Size © 2011 Carolina Biological Supply Company

Ecologists use many different strategies to learn about populations of organisms. While for a few organisms it is possible to simply count them (called "direct census"), many organisms cannot be counted that way. There may be too many, or they may be mobile and move around too much, or they may be difficult to spot. Instead, ecologists take samples of the populations and then use statistical methods to estimate the actual numbers of organisms based on these samples.

In lecture we talk about some ways in which ecologists measure or estimate the size of a population. In this lab you will practice four methods by which populations may be sampled and their sizes estimated.

You will work in small groups. The first day of lab you will complete two exercises, and the second day of lab you will complete two more.

Day 1: Activities A and B

Activity A: Estimating population cover with transects

You should have

- A population sheet (red, blue and green squares)
- A dry erase or overhead marker
- A ruler
- A calculator

You will be using **transects** to sample the populations of blue, red, and green squares on the sheet. A transect is a line through a given area (it may be a string or a stick or rod) along which organisms are counted. This gives the ecologist a sample (a small subset) of the organisms that are present. Multiple transects are run through the area to be sampled, and the results recorded. From the sample data, estimates can be made about the total population in the area. Other types of information that can be gleaned from this kind of sampling include density of a population and diversity of an area.

You are going to be sampling three colors of squares and then making an estimate of the percent coverage for each of the colors for the whole area.

READ THE PROCEDURE INSTRUCTIONS CAREFULLY BEFORE YOU BEGIN! If you have questions, ask me.

Procedure:

1. Using the ruler and the marker, make five lines 10 cm long (these are the transects) on the Population Sheet. Each line should be randomly placed and can be drawn in any direction. All lines must be completely within the colored portion of your sheet.
2. Draw a small dot every 0.5 cm along each transect.

3. Find the 0.0 cm starting point of the first transect and note the color of the square the transect line is touching. Record that color in the "Transect Group Data" table on your data sheet (Transect 1, 0.0 cm).

4. Move along the transect, recording the color of the squares at each 0.5 cm interval in the "Transect Group Data" table. It might be difficult at times to decide which square is touching; be sure you are consistent with however you decide to score the colored squares.

5. Repeat this process for all five transects so that you end up with five complete data sets.

6. Add up the total number of squares of each color. In the "Percent Cover Group Data" table, record these totals in the appropriate column, as well as the overall total number of all the squares you counted.

7. For each color determine the percent cover by **dividing the total number of squares of that color by the overall total squares counted**. Record the percent cover values in the "Percent Cover Group Data" table.

8. Share your percent cover data with the other groups in the class. Record this information in the "Percent Cover Group Data" table.

9. Calculate the **average** for the class data using the data from all the groups. Record the results in the "Percent Cover Group Data" table.

10. Answer the questions for Activity A.

Activity A: Estimating Population Cover with Transects

	Transect Group Data				
	Transect 1	Transect 2	Transect 3	Transect 4	Transect 5
0.0 cm					
0.5 cm					
1.0 cm					
1.5 cm					
2.0 cm					
2.5 cm					
3.0 cm					
3.5 cm					
4.0 cm					
4.5 cm					
5.0 cm					
5.5 cm					
6.0 cm					
6.5 cm					
7.0 cm					
7.5 cm					
8.0 cm					
8.5 cm					
9.0 cm					
9.5 cm					
10.0 cm					

	Percent Cover: Group Data	
	Total	Percent Cover
Red		
Blue		
Green		
Total Squares Counted		

Percent Cover: Class Data									
	Group 1	Group 2	Group 3	Group 4	Group 5	Group 6	Group 7	Group 8	Average
Blue									
Red									
Green									

Activity A: Transect sampling questions

1. In terms of percent cover, using your group's data, rank the colors on the population sheet from highest to lowest.

2. Compare your group's results with the overall class data. Do the results vary from group to group or are they fairly consistent? Describe and explain any inconsistencies among the data sets.

3. How do you think your data might change if you sampled 10 transects rather than only five?

4. In addition to percent cover, what other information might you learn about each population using the transect data you collected?

5. List three situations or types of organisms for which you think this type of population sampling would be appropriate.

6. Give an example of a situation or type of organism in which this strategy would *not* work.

Activity B: Estimating population density using quadrats

You should have

- A population sheet (red, blue and green squares)
- Scissors and clear tape
- A chenille stem
- A ruler
- A calculator

You will be using **quadrats** to sample the population. This is similar to transect sampling but instead of linear samples you get two-dimensional (area) samples. In the field, it is fairly standard to use meter-square quadrats that measure exactly one square meter in area, although quadrats can be any size. In large areas, it is possible to use quadrats that may be 100 m^2 or even a km^2. In these cases sampling is often done from the air, or it can even be done using satellite imagery.

Researchers place the quadrats randomly throughout the area to be sampled and then inventory what they find in each quadrat. When we do our microhabitats lab outdoors in the field, you will use quadrats to sample vegetation at the research site. It will be important to place your quadrats randomly, i.e. not looking for places that seem to have more or fewer plants to census. The statistical probability of getting an accurate result will be low if the sampling is not random.

In our lab today you will make a 6.25-cm^2 quadrat and use it to determine the density of red squares and green squares on the population sheet. In this activity, blue will represent open space. After the density per quadrat for green squares and for red squares is determined, you can then estimate the population size for each color for a given area. Your data could also be used to determine the percent cover for each color, and even the diversity of colors on the population sheet.

READ THE PROCEDURE INSTRUCTIONS CAREFULLY BEFORE YOU BEGIN! If you have questions, ask me.

Procedure:

1. Make a 6.25 cm^2 quadrat using a chenille stem. From the end of the stem, measure 3.0 cm and bend the stem at a right angle. Measure another 3.0 cm along the stem and cut the stem. Repeat these steps to make another right angle; cut the stem. You should now have two right-angles stem pieces. Overlap the stems to create a square with corners that overlap by 0.5 cm. Bend the overlapping ends over one another to secure the stems in the shape of a square. It should measure 2.5 x 2.5 cm.

2. The red and green squares on the population sheet each represent a different species, two species in all. Randomly place the quadrat on the population sheet and secure in place with clear tape.

3. Count the number of red squares and the number of green squares inside the quadrat. It may be difficult to tell for sure around the edges whether a square is inside the quadrat or not; develop a criterion for making this decision and be consistent as you sample the squares.

4. Record the number of green squares and the number of red squares in the table for Activity B. Then remove the quadrat from the population sheet and place it in another location. Repeat the procedure four times, for a total of five (5) samples. When you have five data sets, find the total number of squares of each color for all five samples combined. Record this information in the data table.

5. Calculate the average density of red squares and green squares per 6.25 cm^2 by **dividing the TOTAL number of individuals of each color** (from all five quadrats) **by the total number of quadrats** (5). Record this on the data table.

6. Calculate density per square cm by **dividing the average density of red squares and green squares by the area of one quadrat** (6.25 cm^2). Record this on the data sheet.

7. Answer the questions for Activity B.

Activity B: Estimating Population Density Using Quadrats

Quadrat #	# of red/6.25 cm^2	# of green/6.25 cm^2
1		
2		
3		
4		
5		
Total for 5 quadrats		
Average density/6.25 cm^2		
Density/cm^2		

Activity B: Quadrat sampling questions

1. Given the density calculated for 1 cm^2, what would the estimated population be for an area that is 20 cm x 40 cm in size?

2. Compare your results with those of another group. Describe and explain any inconsistencies between the data sets.

3. Do you think your data might change if you sampled 10 quadrats rather than five? If so, how might it change?

4. In addition to population density, what other information could you determine about each population using the quadrat data you collected?

5. List three situations or types of organisms for which this sampling technique would be appropriate.

6. What is one situation where you can imagine this type of population study would be inappropriate?

Day 2: Activities C and D

Activity C: Estimating population size through mark and recapture

You should have

- A cup of white beads
- A cup of red beads
- 2 small cups
- A calculator

This exercise uses the **mark-and-recapture** method to sample the population of beads in the cup. From your samples, you will estimate the total population of beads in a cup. Scientists in the field will capture and tag or mark a number of individuals of the organism they are studying, and release the organisms back into the habitat. Later the scientists go back and capture another random sample of individuals from the same area. Some of those captured will already be tagged or marked; knowing the percentage of marked individuals recaptured and the original number that were marked, the approximate population size can be calculated.

READ THE PROCEDURE INSTRUCTIONS CAREFULLY BEFORE YOU BEGIN! If you have questions, ask me.

Procedure:

1. The beads in the large cup are the population of organisms found in the research area. White beads are the untagged (or unmarked) individuals and red beads will be the marked individuals. To begin, your entire population is white (unmarked). Start by "capturing" a number of individuals: have one member of the group take a handful of white beads from the cup.

2. Count the white beads you captured and set them aside in a small cup.

3. Now count out the same number of red beads and put them in the large cup with the other uncaptured individuals. You have "marked the "captured" individuals and returned them to the habitat. On your data sheet for Activity C, record the number of red beads (marked individuals) that are now in the population.

4. Stir the population around a little bit so your marked individuals are distributed. Now have a member of the group take another handful of beads from the population. This is the first "recapture" sample and will be recorded as "Sample 1." In the data sheet for Activity C, record how many beads were captured this time. Then, count and record how many of the beads in this sample are red.

5. Calculate the percentage of the sample made up of marked and recaptured individuals. Record this data on the data sheet for Activity C.

6. **Return the sample to the big cup**—all the organisms get released back into the habitat.

7. Repeat steps 4, 5, and 6 until you have captured 20 samples. Calculate the percentage of marked individuals for each sample and record it on the data sheet.

8. Now determine the average percentage of marked and recaptured individuals across all 20 samples. Record this.

9. Estimate the population size.
   a. First multiply the original number of marked individuals by the average number of individuals recaptured.
   b. Then divide that number by the average number of marked individuals recaptured. This is the estimated population size. Record it on the data sheet.

10. Now count the actual number of beads in the cup (now is the time to sort out the colored beads and separate them from the white ones. Thanks!) and see how accurate your estimate was. Do not tell other groups what your answer was before they do the exercise.

11. Answer the questions for Activity C.

Activity C: Estimating population size through mark and recapture

Number of marked individuals (red beads):______________

Sample No.	No. of marked beads recaptured in sample	Total number of beads in sample	% of sample marked and recaptured
1			
2			
3			
4			
5			
6			
7			
8			
9			
10			
11			
12			
13			
14			
15			
16			
17			
18			
19			
20			
Average			

Estimated population size: ________________________________

Actual population size: ________________________________

Activity C: Mark and recapture questions

1. How far off was your estimate from the actual number of beads?

2. Share your results with another group who has completed the exercise. Were your results the same? If not, explain why.

3. How do you think your data might change if you took 40 samples?

4. List three situations or types of organisms for which this sampling technique would be appropriate.

5. Give an example of a situation in which this technique would not work.

Activity D: Estimating population size through removal sampling

You should have

- A tub filled ¾ full of water in which yellow pompom "fish" are "swimming"
- Five small cups
- A net
- A ruler
- A calculator

You will be using **removal sampling** to estimate the population of yellow pompoms in the water. The method graphs the number of individuals captured from a population in successive samples, but (unlike mark and recapture) the individuals are not returned to the habitat. Based on the graph generated by your samples, you will estimate how many pompoms were in the water when you started.

On the graph paper we have, the x-axis will go up to 380, where every 5 boxes=20. The y-axis can be marked in increments of 1, or as necessary based on the first sample.

READ THE PROCEDURE INSTRUCTIONS CAREFULLY BEFORE YOU BEGIN! If you have questions, ask me.

Procedure:

1. You'll need to assign group roles. You need
   a. a collector (this person is in charge of the net and takes pompoms out of the water)
   b. a data recorder (accurately writes down the data from each sample)
   c. an environment manager (this person must see that the water in the tub is swirled at a steady rate, simulating a current, without splashing out or losing any pompoms)

2. Ensure that the same current speed and direction are maintained for each trial (environment manager) and that the net sweeps are performed the same way every time (collector). While the manager swirls the water, the collector should make a single straight sweep through the center of the tub from one end to the other. The idea is to get a random sample of the pompoms.

3. After the sweep, empty the pompoms into one of the cups and have the data recorder count them. Record this number on the "Sampling Trial" data sheet for Trial 1. Then determine the total number of pompoms already removed from the population and enter this number in the table (remember for Trial 1, the total already removed is 0).

4. Repeat the capture, removal, and counting steps four more times for a total of five trials. Make sure the data are entered each time with accuracy.

5. Now graph the data. The "Total Number Removed" data go on the x-axis and the "Number Caught" data go on the y-axis. See note above at the start of this activity for the scale to use when setting up the graph.

6. Using a ruler, determine a best-fit line for the scatter plot you have created on the graph. Draw in this line so it touches both axes of the graph. **The point at which this line touches the x-axis is the estimate of the original population size**.

7. Use the first and last data points to determine the slope of the line (point-slope formula is included on the data sheet). **The slope of your best fit line may not be the same as the calculated slope based on the data points.** This is OK. Show your calculations in the table provided.

8. Now find where the line would cross the x-axis (the x-intercept) based on the slope of the line you calculated. Use **y = mx + b** to find the point at which y = 0 (this is the x-intercept). Show your calculations in the table provided.

9. Answer the questions for Activity D.

Activity D: Estimating population size through removal sampling

	Sampling trial				
	1	2	3	4	5
Number caught					
Total number removed					

Calculations						
y_1	y_2	$y_2 - y_1 = \Delta y$	x_1	x_2	$x_2 - x_1 = \Delta x$	$\Delta y / \Delta x = m$ (slope)

Activity D: Removal sampling questions

1. What do you think are some possible sources of error when using this sampling approach?

2. Compare the results you got to another group's results. Do the results agree? If they don't, tell me why.

3. If you were to take 10 samples rather than five, how do you think the results would change?

4. Give me three situations in which this sampling method would be appropriate.

5. Give me an example of a situation in which this sampling approach would not be useful.

Env Sci 101 Name:___________________________

Lab 7: Microhabitats

A **habitat** is any area in which organisms live. It can be urban, rural, developed or wild. It can be large or on a micro scale. Each organism in a habitat occupies a **niche**, meaning a specific set of ecological parameters. A given space may house many different niches. In this exercise we are going to investigate **microhabitats**. These are areas within a larger space that may vary in surprising ways, including things like temperature, moisture content in soil, exposure to sun, and plant cover. All this variation gives rise to different niches and thus provides habitat for many different types of living things.

Different people in class will take on different roles in this lab. We need 2-3 people per task:

a. **Animal census-takers**. These people will search for animals or signs of animals. This will include laying out a sheet under a shrub nor tree and shaking the shrub or tree and observing and recording what kinds of insects fall onto it (and how many). It will mean turning over rocks and logs and looking underneath them for animal life, and recording those observations.
b. **Plant population samplers.** Using meter squares, these individuals will work to sample three randomly selected plots within each of the designated microhabitats. The plants within each square will be censused, identified and recorded, as will layers of plant cover. Estimated size data will also be recorded for larger plants.
c. **Abiotic data collectors.** Will record observations about abiotic factors within each microhabitat. Data will include information about light availability, soil types, temperatures, drainage, location, exposure to wind, erosion, and slope/aspect.
d. **Soil collectors.** These individuals will be responsible for collecting soil samples (three from each microhabitat) and measuring and recording data on the moisture content of these samples. This is a component of the abiotic data above.

In order to turn in the lab write-up, make sure you gather ALL the data collected by everyone in the class. Fill in all the tables and answer all the questions (below). Then, when all that is finished, answer the following questions. Type up your answers in the form of a report (complete sentences, clear transitions between paragraphs, etc.), including your name, the class, and the date at the top of the page. You should be able to do this in one to two single-spaced pages. Slip the final summary report into the front of this lab manual and turn the entire manual in to me (so I will have ALL the data collected). This will be due on the date given (soil moisture data will take several days to collect).

Questions to address in your report (at a minimum. You may add more):
What differences do you note among the microhabitats in terms of their abiotic components?
What similarities exist across these microhabitats?
What differences exist in terms of plant and animal life among these microhabitats?

Propose an explanation for the differences that you observed. Which factor(s) do you believe have the most influence on the plant and animal life in these microhabitats? Explain your response.

Data for microhabitat A:

1. Description of abiotic features of the microhabitat:

Location:

Relative exposure to light:

Estimated exposure to wind:

Evidence of erosion:

Ground surface temperature (take several measurements and average them):

Slope/aspect:

Approximate dimensions of microhabitat:

Description of drainage:

Soil moisture content (in g): You will need to dig into the soil in three scattered spots within your area of study. Do not include plants from the surface of the soil; these should be replaced when you fill in the hole. Fill a tin about halfway with soil. Then refill the hole as best you can and tamp it down. Label the samples with the microhabitat letter and sample number (A1, A2, A3). Weigh each tin on a spring scale (record which color spring sale you used. *Use the same scale* to measure the dry weight of the soil.), record the mass, and then take the samples back to the classroom to dry out over the next few days.

Complete the table below and then average the percentage water content across the three samples:

Average H_2O content in this plot:

	Initial Mass (g)	Mass after drying (g)	Mass of water (g)	g water/g soil	% water
Sample 1					
Sample 2					
Sample 3					

2. Description of the biotic components of the microhabitat:

Plants

Sample plot 1

Species or type of plant	Number (if countable)	Percent cover (approx.)

Sample plot 2

Species or type of plant	Number (if countable)	Percent cover (approx.)

Sample plot 3

Species or type of plant	Number (if countable)	Percent cover (approx.)

How many layers of vegetation did you identify in this microhabitat?

What was the most abundant plant (or type of plant)? Which type of plant had the greatest coverage?

If this microhabitat has trees or shrubs, describe the estimated height of these plants.

Animals

Describe your sampling methods.

List and describe the types and numbers of animals you observed. Include sketches or photos if applicable.

Data for microhabitat B:

1. Description of abiotic features of the microhabitat:

Location:

Relative exposure to light:

Estimated exposure to wind:

Evidence of erosion:

Ground surface temperature (take several measurements and average them):

Slope/aspect:

Approximate dimensions of microhabitat:

Description of drainage:

Soil moisture content (in g): You will need to dig into the soil in three scattered spots within your area of study. Do not include plants from the surface of the soil; these should be replaced when you fill in the hole. Fill a tin about halfway with soil. Then refill the hole as best you can and tamp it down. Label the samples with the microhabitat letter and sample number (A1, A2, A3). Weigh each tin on a spring scale (record which color spring sale you used. *Use the same scale* to measure the dry weight of the soil.), record the mass, and then take the samples back to the classroom to dry out over the next few days.

Complete the table below and then average the percentage water content across the three samples:

Average H_2O content in this plot:

	Initial Mass (g)	Mass after drying (g)	Mass of water (g)	g water/g soil	% water
Sample 1					
Sample 2					
Sample 3					

2. Description of the biotic components of the microhabitat:

Plants

Sample plot 1

Species or type of plant	Number (if countable)	Percent cover (approx.)

Sample plot 2

Species or type of plant	Number (if countable)	Percent cover (approx.)

Sample plot 3

Species or type of plant	Number (if countable)	Percent cover (approx.)

How many layers of vegetation did you identify in this microhabitat?

What was the most abundant plant (or type of plant)? Which type of plant had the greatest coverage?

If this microhabitat has trees or shrubs, describe the estimated height of these plants.

Animals

Describe your sampling methods.

List and describe the types and numbers of animals you observed. Include sketches or photos if applicable.

Data for microhabitat C:

1. Description of abiotic features of the microhabitat:

Location:

Relative exposure to light:

Estimated exposure to wind:

Evidence of erosion:

Ground surface temperature (take several measurements and average them):

Slope/aspect:

Approximate dimensions of microhabitat:

Description of drainage:

Soil moisture content (in g): You will need to dig into the soil in three scattered spots within your area of study. Do not include plants from the surface of the soil; these should be replaced when you fill in the hole. Fill a tin about halfway with soil. Then refill the hole as best you can and tamp it down. Label the samples with the microhabitat letter and sample number (A1, A2, A3). Weigh each tin on a spring scale (record which color spring sale you used. *Use the same scale* to measure the dry weight of the soil.), record the mass, and then take the samples back to the classroom to dry out over the next few days.

Complete the table below and then average the percentage water content across the three samples:

Average H_2O content in this plot:

	Initial Mass (g)	Mass after drying (g)	Mass of water (g)	g water/g soil	% water
Sample 1					
Sample 2					
Sample 3					

2. Description of the biotic components of the microhabitat:

Plants

Sample plot 1

Species or type of plant	Number (if countable)	Percent cover (approx.)

Sample plot 2

Species or type of plant	Number (if countable)	Percent cover (approx.)

Sample plot 3

Species or type of plant	**Number (if countable)**	**Percent cover (approx.)**

How many layers of vegetation did you identify in this microhabitat?

What was the most abundant plant (or type of plant)? Which type of plant had the greatest coverage?

If this microhabitat has trees or shrubs, describe the estimated height of these plants.

Animals

Describe your sampling methods.

List and describe the types and numbers of animals you observed. Include sketches or photos if applicable.

Env Sci 101 Name________________________________

Lab 8: Invasive Species I

Puereria lobata

Adapted from "Kudz-who? And other questions of invasive species" by D. Parks Collins, Department of Biology, Mitchell Community College, Statesville, NC. © National Center for Case Study Teaching in Science, University at Buffalo, State University of New York. 2017. Used by permission.

As we proceed through the case study, answer the following questions.

1. List several initial thoughts you have after hearing and seeing this brief historical account. Include whether or not you think there was anything wrong with people immediately getting excited about the potential benefits of this particular plant.

2. What are some of the traits that make kudzu particularly invasive?

3. If kudzu was shipped all over the United States, why is it only prevalent in the Southeast? Why did it not establish itself all over the country?

4. List some pros and cons of kudzu below.

Pros________________ Cons______________________

5. Should kudzu have been introduced into the United States?

6. List several initial thoughts you have about the kudzu bug. Does this species provide hope for eradicating kudzu? Why or why not?

7. How would you explain the spread of the kudzu bug? Is it a coincidence that the bug looks to be establishing itself in the same areas as kudzu?

8. List some pros and cons of the "kudzu bug" below.

Pros	Cons

9. Even though the kudzu bug probably was not intentionally introduced, would there be anything wrong with introducing a non-native (or invasive) species to attempt to get rid of another invasive species?

10. If a non-native species in an area seems to not cause any ecological problems, should it still be defined as invasive?

11. Kudzu has been here in the U.S. since 1876. Should it still be considered an invasive species? How do ecologists define a native species? When does a non-native (even if it is invasive) species become a native species? What do we do with native species that are also invasive?

Env Sci 101 Name:______________________________

Lab 9: Invasive Species II
Caulerpa taxifolia

Adapted from "You Poured It Where? A Case Study in Invasive Species" by Nancy M. Boury, Department of Animal Science, Iowa State University. © National Center for Case Study Teaching in Science, University at Buffalo, State University of New York. 2010. Used by permission.

Read the New York Times article provided, "A Delicate Pacific Seaweed Is Now a Monster of the Deep," which talks about a seaweed called *Caulerpa taxifolia.* Also read the story provided.

As we proceed through the story, write down your responses to the following questions:

1. Jim was horrified that Alex planned to dump the water down a storm drain because

2. In your opinion, which single trait of the Aquarium-Mediterranean strain of *Caulerpa taxifolia* makes it particularly desirable for use in public and private saltwater aquariums?

3. What are some of the traits that make *Caulerpa* particularly invasive?

4. An advantage of asexual reproduction over sexual reproduction is

5. The Aquarium-Mediterranean strain of *Caulerpa taxifolia* is particularly invasive because

6. In terms of local ecosystems, why is the invasion of the Mediterranean strain of *Caulerpa taxifolia* a problem?

7. You are in charge of an anti-*Caulerpa* program in a coastal region, what would you do FIRST to control a small patch of the "killer seaweed" found at the bottom of a local fishing harbor?

Env Sci 101 Name:_______________________________

Lab 10: Water use and economic development in Wimberley

Adapted from "To Be or Not To Be a Golf Course in Wimberley?" by Joni Seaton James Charles, Department of Finance and Economics, Texas State University, San Marcos, TX. © the National Center for Case Study Teaching in Science, University at Buffalo, State University of New York. 2014. Used by permission.

First Impressions

Answer the questions below after you take a look at a map of Wimberley, Texas. The first question asks you to think about the town you live in while the second asks you about the town of Wimberley.

1. What are features of the city/town you are living in that make it an attractive place for you to live, go to school, or work in? What are some of the drawbacks? Explain why.

2. General observations are like first impressions. Based on the map you looked at, what are your first impressions of Wimberley and the area surrounding it? Or, how would you describe Wimberley to someone who was considering moving there to live based just on the map?

After learning some more about Wimberley and the Texas Hill Country, answer the next questions.

3. Do you have anything to add to or change about your impressions of Wimberley after reading the information provided?

4. Some of the activities to which visitors are attracted are water dependent, some are not. Which ones are? Which ones are not? How would you measure their value to the town? How should the town measure their value?

5. What are some of the current **challenges** to the natural resources that make Wimberley attractive?

6. Why would the developer choose to apply for a permit to pump water from the underground aquifer given the nature of Wimberley's challenges to its water supply?

7. What are the implications of present versus future benefits and costs? What are the distributional effects of the granted permit?

8. Who are the major stakeholders?

9. How are the different stakeholders likely to view the present benefits and costs of building the golf course versus future benefits and costs?

10. How would one represent this tradeoff between future and present benefits and costs?

11. Why is it important to maintain groundwater quality and quantity? What role did the regulatory authority play?

12. What decision should the city council make? *Why?*

Env Sci 101 Name:_______________________________

Lab 11: PCBs Case Study

Adapted from "PCBs in the Last Frontier: A Case Study on the Scientific Method" by Michael Tessmer Chemistry Department Southwestern College, Winfield, KS. Copyright © 2010 by the National Center for Case Study Teaching in Science. Used by permission.

1. What scientific observation about PCB distribution is described in the introduction?

2. Propose a hypothesis or "explanatory story" to explain the global movement of pollutants such as PCBs. Specifically, how do you think they could end up in the most remote Alaskan lakes?

3. Propose a method, either through observations or direct experimentation, which would test your hypothesis from Question 2 (Note: Your approach may be on a local scale despite examining a global phenomenon.).

4. Come up with a hypothesis or "explanatory story" to answer the following question: Should PCB levels differ significantly in Alaskan lakes that are near each other and at the same latitude? (Keep in mind that a hypothesis is an educated guess, so it requires a reason why you think your answer is correct.)

5. Propose a method, either through observations or direct experimentation, which would test your hypothesis from Question 4.

6. What possible “explanatory story” might explain the observation described above? (Hint: Think of species that leave a lake but return later in life.)

7. How would you test your hypothesis made above?

8. What is the answer to this scientific puzzle? Briefly summarize the findings. Then imagine yourself as a scientist working on this issue. What would you want to look at next?

Env Sci 101 Name:________________________________

Lab 12: Hydropower (part 1): Design phase

Adapted from Carolina STEM Challenge Hydroelectric power © 2016 Carolina Biological Supply Company

We talk in lecture about energy and the various ways society obtains energy to power our civilization. Since the Industrial Revolution, fossil fuels (coal, oil, and gas) have dominated the field of energy. But we know that the combustion of these fuels releases GHGs into the atmosphere, contributing to pollution and global climate change. We also know that these fuels are nonrenewable; that is, once we burn them they are gone and eventually the supply of these fuels from the earth will be exhausted. It took millions and millions of years for the Earth's deposits of carbon-based fuels to form and so they are, from a human standpoint, irreplaceable.

Alternative energy sources are key to maintaining supplies of energy for humans. There are a number of alternative energy technologies, including solar, nuclear, biogas, and hydropower. They all have advantages and disadvantages. In the next two labs we'll explore one of these sources of power, hydropower, and see how the energy in moving water can be used to generate force (power) and do work.

When people set out to solve the kind of problem you will solve in these labs, there are some steps they follow. First you'll need to do your **research**: learn about the problem and any relevant information to help address it. Next you will think about **designs** for solving the problem: what materials will you use? What will your design look like? You will engage in the design phase and build a **prototype**, test it, and **evaluate** it for its success. You will also evaluate other groups' designs using a rubric.

Part 1: Generating power

The hand generator uses a crank to generate power to light up the light bulb.

1. How does the speed of the crank (turning it fast or slowly) affect the light?

2. When a second bulb is added to the generator, what changes? Is there a difference in difficulty in turning the crank? Is there a difference in intensity of the light(s)? Why?

3. What are the similarities between a hand-powered motor and a water- or wind-powered motor?

Part 2: Design Challenge

Goal: You are to design a water wheel that can lift a paper clip when powered by moving water. Your team will demonstrate its design to the class and give a short presentation explaining how and why you designed the wheel the way you did, what problems you faced, and how you might change it if you were to do it over.

The materials are available in the lab to experiment with and work on a design; you will actually build and test the design in the next lab.

You are to design a machine to harness the energy of falling water to turn an axle.

1. Address the following points in your design:

    a. Does the wheel turn smoothly on its axle?
    b. Is it sturdy enough to withstand the force of the falling water without collapsing or falling apart?
    c. How will your design be affected by variations in the flow of water (slowing or speeding up of the falling water)?
    d. How will you actually lift the paper clip?
    e. Are there any safety concerns you will need to be aware of and watch for?

2. Before you leave class today, you must show me a paper from your group that includes the following:

    a. Generate a list of the supplies you will need to build your prototype next lab. If there are materials you do not see in the classroom but would like to have, talk to me about your ideas and we will discuss whether we can get those materials, or if you can bring materials from home.
    b. Sketch and/or describe the design you decide on as a group.
    c. Explain how you will test your design and what factors you will be watching for as you test it. How will you measure your success? What criteria do you want your design to meet?
    d. Get your design approved by me before you leave.

Env Sci 101 Name:________________________________

Lab 12: Hydropower (part 2): Implementation phase

Part 3: Building and testing the prototype

Based on the design from last time, build your water wheel. If you modify the design as you go, make notes about what you had to change and why.

Test it! You'll need to share the water tubs as we only have a few of those. Connect the pipette tip to a length of tubing and connect that tubing to a funnel. Secure all the joints with parafilm. Set up your water wheel.

Someone will need to hold the funnel and another person will need to direct the stream over the wheel. A third person will need to pour the water (carefully!) into the funnel.

Does it work? If you need to modify the design and try it again, do so, and record what you did.

If it works, congratulations! Present the design to the class and demonstrate its operation. Explain what you did and any modifications or alterations you had to make along the way.

Evaluate the designs and presentations of one other group using the rubric provided below.

Finally, answer the questions at the end of the lab.

Evaluation for group____

Criterion	1	2	3	4
Functionality: Does the wheel spin smoothly and consistently under the flow of water?				
Creativity: How creative or novel is the **wheel** design? Unique materials? New approach?				
Effectiveness: Was the water wheel able to smoothly lift the paper clip?				
Creativity: How creative or novel is the **lifting mechanism** design?				
Implementation: Design meets, exceeds, or fails to meet the design criteria				
Testing: Team tested the design and was able to make any modifications as needed				
Presentation: Team's demonstration and explanations were clear and complete				
Applied science: Team demonstrated that they understand the science concepts needed to design this project				

Total score out of 32: ____

Questions:

1. What about your hydropower design worked well?

2. What would you change about your design and why?

3. What are the advantages and disadvantages of using water power? List at least three of each.

12601576R00044

Made in the USA
Monee, IL
27 September 2019